Algebra Practice
1st Edition

equals(me) packets give exciting and challenging entertainment while improving ones mind skills.

equals(me) packets are designed to assist in learning math and algebra concepts. These packets provide excellent algebra practice.

Each packet includes a variety of math and algebra techniques. The packets are modeled using the Study-Try-Compare methods. This gives you the opportunity to study the steps in solving, try for yourself, and compare results. You can also try to solve and then compare your solution.

For school, college, work, or at home, equals(me) packets make math fun.

Disclaimer: The author makes no guarantees as to the accuracy of the material included. This packet is for entertainment purposes only. The included solutions are only that of which the author would perform. There may be several solution methods and/or answers to the equations and/or expressions included in this packet. .

Simplify Expressions

Combine like terms $5x - 3x = 2x$

Use PEMDAS $(3 + 2)(2^2) + 4$

  1. *Parenthesis* $(5)(2^2) + 4$
  2. *Exponents* $(5)(4) + 4$
  3. *Multiply/Divide (left to right)* $20 + 4$
  4. *Add/Subtract (left to right)* 24

Solve equations

Example: $2x - 3 + 3x = x + 5$

Rearrange $2x + 3x - 3 = x + 5$

Combine like terms $5x - 3 = x + 5$

Isolate variable by moving variable $5x - x = 5 + 3$
to left of equal sign and everything
else to right of equal sign. *

Combine like terms again. $4x = 8$

Isolate variable again using $x = \frac{8}{4}$
inverse operations. *
                                                $x = 2$

* *Notes:*
If they are added to the one side, subtract on the other side.
If they are subtracted to the one side, add to the other side.
If they are multiplied to the one side, divide on the other side.
If they are divided on the one side, multiply to the other side.

Inequalities

Combine like terms	$3x - 6x = -3x$

Reverse the inequality when multiplying or dividing with negative divisor

$$-2y < -8$$
$$y > \frac{-8}{-2}$$
$$y > 4$$

Reverse the inequality when swapping sides

$$2 + 12 < 3x$$
$$3x > 2 + 12$$

Change a numbers sign when moving it to other side

$$3x + 4 = -2x - 3$$
$$3x + 2x = -3 - 4$$

Get variable alone

$$x \geq 3 - 5$$

Factoring

Find roots of expression

$$x^2 + 5x + 6 = (x + 2)(x + 3)$$

Find 2 numbers for which their sum will equal the 2nd term, and their product will equal the 3rd term.

Use FOIL to check:
Combining like terms:

$$(x + 2)(x + 3)$$
$$x^2 + 2x + 3x + 6$$
$$x^2 + 5x + 6$$

See pages 78 & 79 for more rules of algebra.

Simplify Expressions

$5(x - 8) + 7(x + 5)$

$5x - 40 + 7x + 35$

$5x + 7x - 40 + 35$

$12x - 5$

$9(x + 7) - 4(-x + 3)$

$9x + 63 + 4x - 12$

$9x + 4x + 63 - 12$

$13x + 51$

$7(-x - 3)(x + 4)$

$7(-x^2 - 4x - 3x - 12)$

$7(-x^2 - 7x - 12)$

$-7x^2 - 49x - 84$

$4(x - 5)(3x + 8)(2 - 5)$

$(4x - 20)(3x + 8)(-3)$

$(4x - 20)(-9x - 24)$

$-36x^2 - 96x + 180x + 480$

$-36x^2 + 84x + 480$

Simplify Expressions

$$5(x - 8) + 7(x + 5)$$

$$9(x + 7) - 4(-x + 3)$$

$$7(-x - 3)(x + 4)$$

$$4(x - 5)(3x + 8)(2 - 5)$$

Simplify Expressions

$(6x - 3) + (4x + 8) - 12$

$6x + 4x - 3 + 8 - 12$

$10x - 7$

$2(4x - 7) - (12x + 5) - 14$

$8x - 14 -12x - 5 - 14$

$8x - 12x - 14 - 5 - 14$

$-4x -33$

$(x - 4) - 7x(9 - 12) + 21$

$(x - 4) - 7x(-3) + 21$

$x - 4 + 21x + 21$

$x + 21x - 4 + 21$

$22x + 17$

$y(5 - 12) - 4y(-15 + 3)$

$y(-7) - 4y(-12)$

$-7y + 48y$

$41y$

Simplify Expressions

$$(6x - 3) + (4x + 8) - 12$$

_____

_____

_____

_____

_____

$$2(4x - 7) - (12x + 5) - 14$$

_____

_____

_____

_____

$$(x - 4) - 7x(9 - 12) + 21$$

_____

_____

_____

_____

$$y(5 - 12) - 4y(-15 + 3)$$

_____

_____

_____

_____

Simplify Expressions

$3 - (-5y - 7)$

$5y + 3 - -7$

$5y + 3 + 7$

$5y + 10$

$-10 + 5(6x - 4)$

$-10 + 30x - 20$

$30x - 20 - 10$

$30x - 30$

$5(-7y - 8) - 4(10 + 2y)$

$-35y - 40 - 40 - 8y$

$-35y - 8y - 40 - 40$

$-43y - 80$

$(-3a - 5) + 9(-3 + 4a)$

$(-3a - 5) - 27 + 36a$

$-3a + 36a - 5 - 27$

$33a - 32$

Simplify Expressions

$$3 - (-5y - 7)$$

$$-10 + 5(6x - 4)$$

$$5(-7y - 8) - 4(10 + 2y)$$

$$(-3a - 5) + 9(-3 + 4a)$$

Simplify Expressions

5(-3x - 9) - 2(4x - 9)

-15x - 45 - 8x + 18

-15x - 8x + 18 - 45

-23x - 27

-4x(-7) - 2(-3x - 9)

28x + 6x + 18

34x + 18

9(x + 8) + 3(12x - 7)

9x + 72 + 36x - 21

9x + 36x + 72 - 21

45x + 51

7(8x - 7) - 7(-6x - 9)

56x - 49 + 42x + 63

56x + 42x + 63 - 49

98x + 14

Simplify Expressions

$$5(-3x - 9) - 2(4x - 9)$$

$$-4x(-7) - 2(-3x - 9)$$

$$9(x + 8) + 3(12x - 7)$$

$$7(8x - 7) - 7(-6x - 9)$$

Simplify Expressions

5(2x + 3) + (x - 2) - 8

```
10x + 15 + (x - 2) - 8
10x + x + 15 - 2 - 8
11x + 5
```

3(x - 5)(2) - (4x + 7)

```
(3x - 15)(2) - 4x - 7
6x - 30 - 4x - 7
6x - 4x - 30 - 7
2x - 37
```

(2x - 5x)(4) + 3(x - 3)(2) + 5

```
8x - 20x + (3x - 9)(2) + 5
8x - 20x + 6x - 18 + 5
-6x - 13
```

---

x(x - 3) - x(2x + 5)

$$x^2 - 3x - 2x^2 - 5x$$
$$-2x^2 + x^2 - 3x - 5x$$
$$-1x^2 - 8x$$
$$-x^2 - 8x$$

Simplify Expressions

$$5(2x + 3) + (x - 2) - 8$$

_____

_____

_____

_____

_____

$$3(x - 5)(2) - (4x + 7)$$

_____

_____

_____

_____

$$(2x - 5x)(4) + 3(x - 3)(2) + 5$$

_____

_____

_____

_____

$$x(x - 3) - x(2x + 5)$$

_____

_____

_____

_____

Simplify Expressions

$$\left(\tfrac{x}{5}\right)\left(\tfrac{30}{x}\right) + 12$$

$$\tfrac{30x}{5x} + 12$$

$$6 + 12$$

$$18$$

$$\left(\tfrac{x}{5}\right)(5) + 4x$$

$$\left(\tfrac{x}{5}\right)\left(\tfrac{5}{1}\right) + 4x$$

$$\left(\tfrac{5x}{5}\right) + 4x$$

$$x + 4x$$

$$5x$$

$$\left(\tfrac{x}{4}\right)(-12) - 7x + 5$$

$$\left(\tfrac{x}{4}\right)\left(\tfrac{-12}{1}\right) - 7x + 5$$

$$\tfrac{-12x}{4} - 7x + 5$$

$$-3x - 7x + 5$$

$$-10x + 5$$

$$-4x\left(\tfrac{x}{4}\right) - 20x$$

$$\left(-\tfrac{4x}{1}\right)\left(\tfrac{x}{4}\right) - 20x$$

$$\left(-\tfrac{4x^2}{4}\right) - 20x$$

$$-x^2 - 20x$$

Simplify Expressions

$$\left(\frac{x}{5}\right)\left(\frac{30}{x}\right) + 12$$

_____
_____
_____
_____
_____

$$\left(\frac{x}{5}\right)(5) + 4x$$

_____
_____
_____
_____
_____

$$\left(\frac{x}{4}\right)(-12) - 7x + 5$$

_____
_____
_____
_____

$$-4x\left(\frac{x}{4}\right) - 20x$$

_____
_____
_____
_____

Simplify Expressions

$$\frac{14}{7} - 5x + 2(\tfrac{1}{2}) - 7$$

$$2 - 5x + 1 - 7$$

$$-5x + 2 + 1 - 7$$

$$-5x - 4$$

$$2x^2 + 2(x - 5)(x + 5) - 12$$

$$2x^2 + 2(x^2 + 5x - 5x - 25) - 12$$

$$2x^2 + 2(x^2 - 25) - 12$$

$$2x^2 + 2x^2 - 50 - 12$$

$$4x^2 - 62$$

$$3(x - 8)(x + 5) - 32$$

$$3(x^2 + 5x - 8x - 40) - 32$$

$$3x^2 + 15x - 24x - 120 - 32$$

$$3x^2 - 9x - 152$$

$$4(\tfrac{3}{4}) + 3x - 8(x - 2)$$

$$3 + 3x - 8x + 16$$

$$3x - 8x + 16 + 3$$

$$-5x + 19$$

Simplify Expressions

$$\frac{14}{7} - 5x + 2\left(\frac{1}{2}\right) - 7$$

$$2x^2 + 2(x - 5)(x + 5) - 12$$

$$3(x - 8)(x + 5) - 32$$

$$4\left(\frac{3}{4}\right) + 3x - 8(x - 2)$$

Simplify Expressions

5 - 7(x - 7x) - 2(x - 5) - 4

$5 - 7x + 49x - 2x + 10 - 4$

$49x - 7x - 2x + 10 + 5 - 4$

$40x + 11$

(x - 9)(5) + 2x - 8

$5x - 45 + 2x - 8$

$5x + 2x - 45 - 8$

$7x - 53$

2(x - 3)(x + 5)(4)

$(2x - 6)(x + 5)(4)$

$(2x - 6)(4x + 20)$

$8x^2 + 40x - 24x - 120$

$8x^2 + 16x - 120$

3x(3x + 5)(8 - 3) - 20x + 8

$(9x^2 + 15x)(5) - 20x + 8$

$45x^2 + 75x - 20x + 8$

$45x^2 + 55x + 8$

Simplify Expressions

$$5 - 7(x - 7x) - 2(x - 5) - 4$$

_____

_____

_____

_____

_____

$$(x - 9)(5) + 2x - 8$$

_____

_____

_____

_____

$$2(x - 3)(x + 5)(4)$$

_____

_____

_____

_____

$$3x(3x + 5)(8 - 3) - 20x + 8$$

_____

_____

_____

_____

Simplify Expressions

$2x^2 + 3x - 10x^2 - 7x$

$2x^2 - 10x^2 + 3x - 7x$

$-8x^2 - 4x$

$(5)(5x^2)(-5x)$

$(5)(-25x^3)$

$-125x^3$

$-1 - y^2 - 3y^2$

$-3y^2 - y^2 - 1$

$-4y^2 - 1$

$(-x)(\frac{7x^2}{x})$

$(\frac{-x}{1})(\frac{7x^2}{x})$

$\frac{-7x^3}{x}$

$-7x^2$

Simplify Expressions

$$2x^2 + 3x - 10x^2 - 7x$$

_____
_____
_____
_____
_____

$$(5)(5x^2)(-5x)$$

_____
_____
_____
_____

$$-1 - y^2 - 3y^2$$

_____
_____
_____
_____

$$(-x)\left(\frac{7x^2}{x}\right)$$

_____
_____
_____
_____

Simplify Expressions

$5(2 + x) + 3(5x + 4) - (x^2)^2$

$10 + 5x + 15x + 12 - x^4$

$-x^4 + 5x + 15x + 10 + 12$

$- x^4 + 20x + 22$

$3(5 - x) + 4(5x - 3) + (x^2)^3$

$15 - 3x + 20x - 12 + x^6$

$x^6 - 3x + 20x + 15 - 12$

$x^6 + 17x + 3$

$4(x - 5) - 3(2x - 8) + x^2$

$4x - 20 - 6x + 24 + x^2$

$x^2 + 4x - 6x + 24 - 20$

$x^2 - 2x + 4$

$2(x - 3) + 4b - 2(x - b - 3) + 5$

$2x - 6 + 4b - 2x + 2b + 6 + 5$

$2x - 2x + 4b + 2b - 6 + 6 + 5$

$6b + 5$

Simplify Expressions

$$5(2 + x) + 3(5x + 4) - (x^2)^2$$

_____

_____

_____

_____

_____

$$3(5 - x) + 4(5x - 3) + (x^2)^3$$

_____

_____

_____

_____

$$4(x - 5) - 3(2x - 8) + x^2$$

_____

_____

_____

_____

$$2(x - 3) + 4b - 2(x - b - 3) + 5$$

_____

_____

_____

_____

Inequalities

$$9x - 3 \geq 8 - 2x$$

$$
\begin{aligned}
9x - 3 + 2x &\geq 8 \\
9x + 2x &\geq 8 + 3 \\
11x &\geq 11 \\
x &\geq \frac{11}{11} \\
x &\geq 1
\end{aligned}
$$

$$12y - 4 \geq 8 - 6y$$

$$
\begin{aligned}
12y - 4 + 6y &\geq 8 \\
12y + 6y &\geq 8 + 4 \\
18y &\geq 12 \\
y &\geq \frac{12}{18} \\
y &\geq \frac{2}{3}
\end{aligned}
$$

$$5x - 12 < 4x + 8$$

$$
\begin{aligned}
5x - 12 - 4x &< 8 \\
5x - 4x &< 8 + 12 \\
1x &< 20 \\
x &< \frac{20}{1} \\
x &< 20
\end{aligned}
$$

$$x - 18 \leq 2x + 3$$

$$
\begin{aligned}
x - 18 - 2x &\leq 3 \\
x - 2x &\leq 3 + 18 \\
-1x &\leq 21 \\
x &\geq \frac{21}{-1} \\
x &\geq -21
\end{aligned}
$$

Inequalities

$$9x - 3 \geq 8 - 2x$$

$$12y - 4 \geq 8 - 6y$$

$$5x - 12 < 4x + 8$$

$$x - 18 \leq 2x + 3$$

Inequalities

$$14x > 4x + 15$$

$$14x - 4x > 15$$
$$10x > 15$$
$$x > \frac{15}{10}$$
$$x > \frac{3}{2}$$
$$x > 1\tfrac{1}{2}$$

$$14 - 3x < 12 + 7x$$

$$14 - 3x - 7x < 12$$
$$-3x - 7x < 12 - 14$$
$$-10x < -2$$
$$x > \frac{-2}{-10}$$
$$x > \frac{1}{5}$$

$$7x - 20 \geq 7x + 2$$

$$7x - 20 - 7x \geq 2$$
$$7x - 7x \geq 2 + 20$$
$$0 \geq 22$$
$$No\ real\ solutions$$

$$3x - 1 \leq 5x + 1$$

$$3x - 1 - 5x \leq 1$$
$$3x - 5x \leq 1 + 1$$
$$-2x \leq 2$$
$$x \geq \frac{2}{-2}$$
$$x \geq -1$$

Inequalities

$$14x > 4x + 15$$

_____

$$14 - 3x < 12 + 7x$$

_____

$$7x - 20 \geq 7x + 2$$

_____

$$3x - 1 \leq 5x + 1$$

_____

Inequalities

$$6(13 - 5b) - (b + 11) \geq 36$$

$$78 - 30b - b - 11 \geq 36$$
$$-31b + 67 \geq 36$$
$$-31b \geq 36 - 67$$
$$-31b \geq -31$$
$$b \leq \frac{-31}{-31}$$
$$b \leq 1$$

$$4 + 5c \geq 2c + 4$$

$$5c - 2c \geq 4 - 4$$
$$3c \geq 0$$
$$c \geq \frac{0}{3}$$
$$c \geq 0$$

$$3b - 2 + 5b + 10 < -5(6b - 8) + 4(9b - 12)$$

$$8b + 8 < -30b + 40 + 36b - 48$$
$$8b + 8 < 6b - 8$$
$$8b - 6b < -8 - 8$$
$$2b < -16$$
$$b < \frac{-16}{2}$$
$$b < -8$$

$$10 + a \leq 18 + 6a - 3a$$

$$10 + a \leq 18 + 3a$$
$$a - 3a \leq 18 - 10$$
$$-2a \leq 8$$
$$a \geq \frac{8}{-2}$$
$$a \geq -4$$

Inequalities

$$6(13 - 5b) - (b + 11) \geq 36$$

_____
_____
_____
_____
_____
_____

$$4 + 5c \geq 2c + 4$$

_____
_____
_____
_____

$$3b - 2 + 5b + 10 < -5(6b - 8) + 4(9b - 12)$$

_____
_____
_____
_____
_____

$$10 + a \leq 18 + 6a - 3a$$

_____
_____
_____
_____

Inequalities

$$6 + 4b \geq 2b + 8$$
$$4b - 2b \geq 8 - 6$$
$$2b \geq 2$$
$$b \geq \frac{2}{2}$$
$$b \geq 1$$

$$-x + 7x \leq -2 + 7x$$
$$-x + 7x - 7x \leq -2$$
$$-x \leq -2$$
$$x \geq \frac{-2}{-1}$$
$$x \geq 2$$

$$8 + 7x \geq 4x + 14$$
$$7x - 4x \geq 14 - 8$$
$$3x \geq 6$$
$$x \geq \frac{6}{3}$$
$$x \geq 2$$

$$-6x + 5x \leq 8$$
$$-x \leq 8$$
$$x \geq \frac{8}{-1}$$
$$x \geq -8$$

Inequalities

$$6 + 4b \geq 2b + 8$$

$$-x + 7x \leq -2 + 7x$$

$$8 + 7x \geq 4x + 14$$

$$-6x + 5x \leq 8$$

Inequalities

$$17 \leq \tfrac{x}{2} + 11$$

$$2 * 17 \leq 2(\tfrac{x}{2} + 11)$$

$$34 \leq x + 22$$

$$34 - 22 \leq x$$

$$12 \leq x$$

$$x \geq 12$$

$$-7 \leq \tfrac{x+2}{4}$$

$$4(-7) \leq 4(\tfrac{x+2}{4})$$

$$-28 \leq x + 2$$

$$-28 - 2 \leq x$$

$$-30 \leq x$$

$$x \geq -30$$

$$-9 < \tfrac{x}{5} - 2$$

$$5(-9) < 5(\tfrac{x}{5} - 2)$$

$$-45 < x - 10$$

$$-x > -10 + 45$$

$$-x > 35$$

$$x > -35$$

$$-6 + \tfrac{x}{3} \leq 8 + 3$$

$$3(-6 + \tfrac{x}{3}) \leq 3(8 + 3)$$

$$-18 + x \leq 24 + 9$$

$$x \leq 24 + 9 + 18$$

$$x \leq 51$$

Inequalities

$$17 \leq \tfrac{x}{2} + 11$$

$$-7 \leq \tfrac{x+2}{4}$$

$$-9 < \tfrac{x}{5} - 2$$

$$-6 + \tfrac{x}{3} \leq 8 + 3$$

Inequalities

$$-2(9 - 3x) - (8 - x) \geq x(24 - 4)$$
$$-18 + 6x - 8 + x \geq 20x$$
$$6x + x - 20x \geq 18 + 8$$
$$-13x \geq 26$$
$$x \leq \frac{26}{-13}$$
$$x \leq -2$$

$$2x(15 - 3) < (x + 3) - (4x - 2)$$
$$24x < x - 4x + 3 + 2$$
$$24x < -3x + 5$$
$$24x + 3x < 5$$
$$27x < 5$$
$$x < \frac{5}{27}$$

$$-4y + 5y > 21$$
$$1y > 21$$
$$y > 21$$

$$-4y - 5y > 3(8 + y)$$
$$-9y > 24 + 3y$$
$$-9y - 3y > 24$$
$$-12y > 24$$
$$y < \frac{24}{-12}$$
$$y < -2$$

Inequalities

$$-2(9 - 3x) - (8 - x) \geq x(24 - 4)$$

$$2x(15 - 3) < (x + 3) - (4x - 2)$$

$$-4y + 5y > 21$$

$$-4y - 5y > 3(8 + y)$$

Inequalities

$$-8 < 9x + 7 < -2$$
$$-8 - 7 < 9x < -2 - 7$$
$$-15 < 9x < -9$$
$$\frac{-15}{9} < x < \frac{-9}{9}$$
$$\frac{-5}{3} < x < -1$$

$$1 < 2x - 10 < 3$$
$$1 + 10 < 2x < 3 + 10$$
$$11 < 2x < 13$$
$$\frac{11}{2} < x < \frac{13}{2}$$

$$-4 < 3x - 2 < 8$$
$$-4 + 2 < 3x < 8 + 2$$
$$-2 < 3x < 10$$
$$\frac{-2}{3} < x < \frac{10}{3}$$

$$-10 < 5x - 7 < -4$$
$$-10 + 7 < 5x < -4 + 7$$
$$-3 < 5x < 3$$
$$\frac{-3}{5} < x < \frac{3}{5}$$

Inequalities

$$-8 < 9x + 7 < -2$$

$$1 < 2x - 10 < 3$$

$$-4 < 3x - 2 < 8$$

$$-10 < 5x - 7 < -4$$

Solve for variable

$$12 + 10(x - 2) - 7 = 95$$

$$12 + 10x - 20 - 7 = 95$$
$$10x - 15 = 95$$
$$10x = 95 + 15$$
$$10x = 110$$
$$x = \frac{110}{10}$$
$$x = 11$$

$$\frac{x}{4} + (14 - 2) = 48$$

$$\frac{x}{4} + 12 = 48$$
$$\frac{x}{4} = 48 - 12$$
$$\frac{x}{4} = 36$$
$$x = 36 * 4$$
$$x = 144$$

$$8x + 20 - (6 + 2) = 60$$

$$8x + 12 = 60$$
$$8x = 60 - 12$$
$$8x = 48$$
$$x = \frac{48}{8}$$
$$x = 6$$

$$\frac{x}{3} * 5 = 25$$

$$\frac{x}{3} = \frac{25}{5}$$
$$\frac{x}{3} = 5$$
$$x = 5 * 3$$
$$x = 15$$

Solve for variable

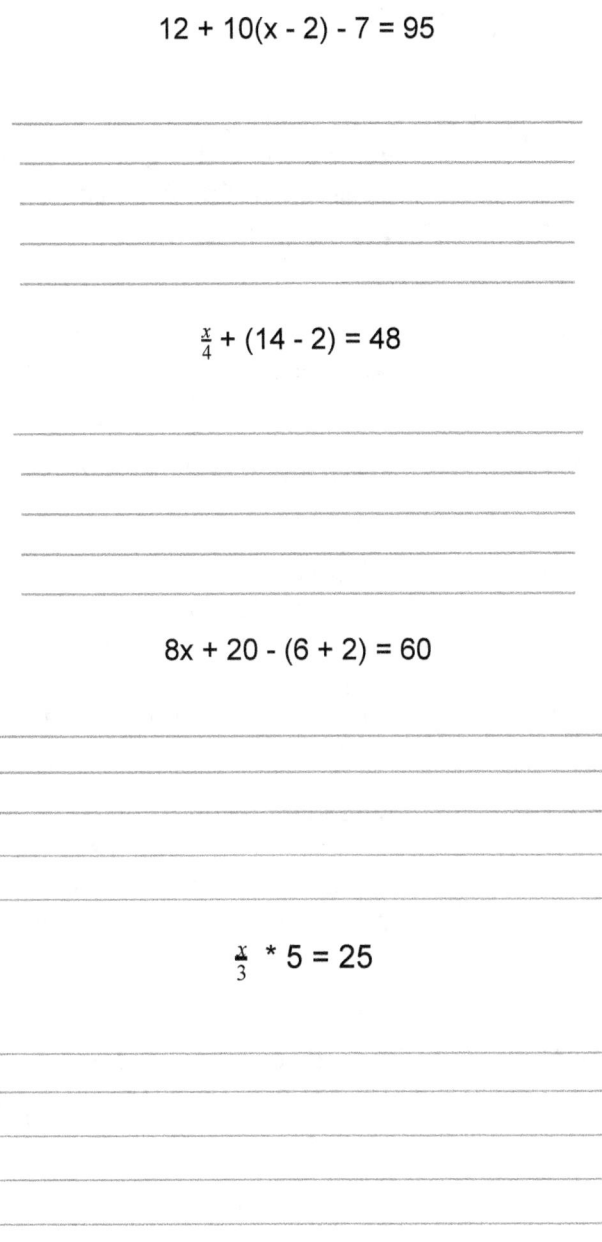

$$12 + 10(x - 2) - 7 = 95$$

_____
_____
_____
_____
_____

$$\frac{x}{4} + (14 - 2) = 48$$

_____
_____
_____
_____

$$8x + 20 - (6 + 2) = 60$$

_____
_____
_____
_____

$$\frac{x}{3} * 5 = 25$$

_____
_____
_____
_____

Algebra Definitions

Expressions	A mathematical phrase that may contain numbers, variables, or operations, but does not include a relationship symbol such as =, <, or >.
Variables	A letter used that represents a quantity that can change.
Coefficient	The number part of a term with a variable.
Constants	A term containing only numbers. Constants do not include variables.
Binomial	A polynomial with two terms.
Polynomial	An expression consisting of variables and coefficients that involves only the operations of addition, subtraction, multiplication, and non-negative integer exponents.
Inequality	A sentence that states one expression is greater than, less than or equal to another expression

Did you know......

The number Pi can't be expressed as a fraction, making it an irrational number. It never repeats and never ends when written as a decimal.

Algebra is used in fields like science, engineering, economics, mathematics and medicine.

If you divide 1 by a number, and divide 1 by that number, you end up with the number you started with.

Zero is the only number that cannot be represented with roman numerals.

In Thailand, 5 is pronounced 'ha'. So 555 is 'hahaha'.

Albert Einstein did not speak until he was 4 years old.

Benjamin Franklin once said "No employment can be managed without arithmetic, no mechanical invention without geometry."

The equals sign (=) was invented in 1557 by a mathematician named Robert Recorde.

Solve for variable

$$-14 + 5(x - 3)(2 + 4) = 16$$

$$-14 + (5x - 15)(6) = 16$$
$$-14 + 30x - 90 = 16$$
$$30x - 104 = 16$$
$$30x = 16 + 104$$
$$30x = 120$$
$$x = \frac{120}{30}$$
$$x = 4$$

$$15 - 3(x + 2)(3 - 5) = 24$$

$$15 - (3x - 6)(-2) = 24$$
$$15 + 6x + 12 = 24$$
$$6x + 27 = 24$$
$$6x = 24 - 27$$
$$6x = -3$$
$$x = \frac{-3}{6}$$
$$x = \frac{-1}{2}$$

$$8 - 5(x - 8)(6 - 2) = 2(x - 4)$$

$$8 - (5x + 40)(4) = 2x - 8$$
$$8 - 20x + 160 = 2x - 8$$
$$-20x + 168 = 2x - 8$$
$$-20x - 2x = -8 - 168$$
$$-22x = -176$$
$$x = \frac{-176}{-22}$$
$$x = 8$$

Solve for variable

$$-14 + 5(x - 3)(2 + 4) = 16$$

$$15 - 3(x + 2)(3 - 5) = 24$$

$$8 - 5(x - 8)(6 - 2) = 2(x - 4)$$

Solve for variable

$$2(x + 5) + 3 = 3(x + 3)$$

$$2x + 10 + 3 = 3x + 9$$

$$2x - 3x = 9 - 10 - 3$$

$$-x = -4$$

$$x = \frac{-4}{-1}$$

$$x = 4$$

$$2(x + 5) = 2(x + 3) + 4$$

$$2x + 10 = 2x + 6 + 4$$

$$2x - 2x = 6 + 4 - 10$$

$$0 = 0$$

All real numbers are solutions

$$3(x - 2) = 2(x + 3)$$

$$3x - 6 = 2x + 6$$

$$3x - 2x = 6 + 6$$

$$x = 12$$

$$4(x - 2) = 2(x - 4)$$

$$4x - 8 = 2x - 8$$

$$4x - 2x = -8 + 8$$

$$2x = 0$$

$$x = \frac{0}{2}$$

$$x = 0$$

Solve for variable

$$2(x + 5) + 3 = 3(x + 3)$$

$$2(x + 5) = 2(x + 3) + 4$$

$$3(x - 2) = 2(x + 3)$$

$$4(x - 2) = 2(x - 4)$$

Solve for variable

$$6x + 5 = 3x + 14$$
$$6x - 3x = 14 - 5$$
$$3x = 9$$
$$x = \frac{9}{3}$$
$$x = 3$$

$$5y + 9 = 2y + 18$$
$$5y - 2y = 18 - 9$$
$$3y = 9$$
$$y = \frac{9}{3}$$
$$y = 3$$

$$3a + 4 = a + 22$$
$$3a - a = 22 - 4$$
$$2a = 18$$
$$a = \frac{18}{2}$$
$$a = 9$$

$$3y + 31 = 8y + 6$$
$$3y - 8y = 6 - 31$$
$$-5y = -25$$
$$y = \frac{-25}{-5}$$
$$y = 5$$

Solve for variable

$$6x + 5 = 3x + 14$$

_____
_____
_____
_____
_____

$$5y + 9 = 2y + 18$$

_____
_____
_____
_____
_____

$$3a + 4 = a + 22$$

_____
_____
_____
_____

$$3y + 31 = 8y + 6$$

_____
_____
_____
_____
_____

Solve for variable

$$2(\tfrac{x}{2}) + 7 = 3x - 5$$

$$\tfrac{2x}{2} + 7 = 3x - 5$$

$$x + 7 = 3x - 5$$

$$x - 3x = -5 - 7$$

$$-2x = -12$$

$$x = \tfrac{-12}{-2}$$

$$x = 6$$

$$3(\tfrac{x}{3}) - 5 = x - 5$$

$$\tfrac{3x}{3} - 5 = x - 5$$

$$x - 5 = x - 5$$

$$x - x = -5 + 5$$

$$0 = 0$$

All real numbers are solutions

$$7y - 8y + 3 = 4 - 9$$

$$7y - 8y = 4 - 9 - 3$$

$$-1y = -8$$

$$y = \tfrac{-8}{-1}$$

$$y = 8$$

$$2(x + 2)(3) = 2x - 8$$

$$(2x + 4)(3) = 2x - 8$$

$$6x + 12 = 2x - 8$$

$$6x - 2x = -8 - 12$$

$$4x = -20$$

$$x = \tfrac{-20}{4}$$

$$x = -5$$

Solve for variable

$$2\left(\frac{x}{2}\right) + 7 = 3x - 5$$

$$3\left(\frac{x}{3}\right) - 5 = x - 5$$

$$7y - 8y + 3 = 4 - 9$$

$$2(x + 2)(3) = 2x - 8$$

Solve for variable

$$10x + 1 = x + 10$$
$$10x - x + 1 = 10$$
$$10x - x = 10 - 1$$
$$9x = 9$$
$$x = \frac{9}{9}$$
$$x = 1$$

$$-7x - 8 = -5x - 6$$
$$-7x + 5x - 8 = -6$$
$$-7x + 5x = -6 + 8$$
$$-2x = 2$$
$$x = \frac{2}{-2}$$
$$x = -1$$

$$-8x - 1 = -2x - 4$$
$$-8x + 2x - 1 = -4$$
$$-8x + 2x = -4 + 1$$
$$-6x = -3$$
$$x = \frac{-3}{-6}$$
$$x = \frac{1}{2}$$

$$2(x + 2) = 4(x - 2)$$
$$2x + 4 = 4x - 8$$
$$2x - 4x + 4 = -8$$
$$2x - 4x = -8 - 4$$
$$-2x = -12$$
$$x = \frac{-12}{-2}$$
$$x = 6$$

Solve for variable

$$10x + 1 = x + 10$$

$$-7x - 8 = -5x - 6$$

$$-8x - 1 = -2x - 4$$

$$2(x + 2) = 4(x - 2)$$

Solve for variable

$$\left(\tfrac{1}{4}\right)x + \left(\tfrac{1}{3}\right)x = 21$$

$$\left(\tfrac{3}{12}\right)x + \left(\tfrac{4}{12}\right)x = 21$$

$$12\left(\tfrac{3}{12}\right)x + 12\left(\tfrac{4}{12}\right)x = (12)(21)$$

$$3x + 4x = 252$$

$$7x = 252$$

$$x = \tfrac{252}{7}$$

$$x = 36$$

$$\left(\tfrac{1}{4}\right)y + \left(\tfrac{1}{2}\right)y = -18$$

$$4\left(\tfrac{1}{4}\right)y + 4\left(\tfrac{1}{2}\right)y = (4)(-18)$$

$$1y + 2y = -72$$

$$3y = -72$$

$$y = \tfrac{-72}{-3}$$

$$y = -24$$

$$\left(\tfrac{1}{2}\right)b + \left(\tfrac{1}{4}\right)b = -12$$

$$4\left(\tfrac{1}{2}\right)b + 4\left(\tfrac{1}{4}\right)b = (4)(-12)$$

$$2b + 1b = -48$$

$$3b = -48$$

$$b = \tfrac{-48}{3}$$

$$b = -16$$

$$\left(\tfrac{1}{5}\right)x + \left(\tfrac{1}{3}\right)x = -40$$

$$\left(\tfrac{3}{15}\right)x + \left(\tfrac{5}{15}\right)x = -40$$

$$15\left(\tfrac{3}{15}\right)x + 15\left(\tfrac{5}{15}\right)x = (15)(-40)$$

$$3x + 5x = -600$$

$$8x = -600$$

$$x = \tfrac{-600}{8}$$

$$x = -75$$

Solve for variable

$$\left(\tfrac{1}{4}\right)x + \left(\tfrac{1}{3}\right)x = 21$$

$$\left(\tfrac{1}{4}\right)y + \left(\tfrac{1}{2}\right)y = -18$$

$$\left(\tfrac{1}{2}\right)b + \left(\tfrac{1}{4}\right)b = -12$$

$$\left(\tfrac{1}{5}\right)x + \left(\tfrac{1}{3}\right)x = -40$$

Factoring - Solve for x

$$x^2 + 12x = -20$$

$x^2 + 12x + 20 = 0$

$(x + 2)(x + 10) = 0$

Set factors to equal 0

$(-2 + 2 = 0)$ & $(-10 + 10 = 0)$

$x = -2$ or $x = -10$

$$x^2 + 11x = -18$$

$x^2 + 11x + 18 = 0$

$(x + 2)(x + 9) = 0$

Set factors to equal 0

$(-2 + 2 = 0)$ & $(-9 + 9 = 0)$

$x = -2$ or $x = -9$

$$x^2 + 6x = 16$$

$x^2 + 6x - 16 = 0$

$(x - 2)(x + 8) = 0$

Set factors to equal 0

$(2 - 2 = 0)$ & $(-8 + 8 = 0)$

$x = 2$ or $x = -8$

Factoring - Solve for x

$$x^2 + 12x = -20$$

$$x^2 + 11x = -18$$

$$x^2 + 6x = 16$$

Factoring - Solve for x

$$x^2 + 4x = 21$$

$$x^2 + 4x - 21 = 0$$
$$(x - 3)(x + 7) = 0$$
Set factors to equal 0
$$(3 - 3 = 0) \quad \& \quad (-7 + 7 = 0)$$
$$x = 3 \text{ or } x = -7$$

$$2x^2 - 12x = -18$$

$$2x^2 - 12x + 18 = 0$$
$$2(x - 3)(x - 3) = 0$$
Set factors to equal 0
$$(3 - 3 = 0) \quad \& \quad (3 - 3 = 0)$$
$$x = 3$$

$$2x^2 - 32x = -128$$

$$2x^2 - 32x + 128 = 0$$
$$2(x - 8)(x - 8) = 0$$
Set factors to equal 0
$$(8 - 8 = 0) \quad \& \quad (8 - 8 = 0)$$
$$x = 8$$

Factoring - Solve for x

$$x^2 + 4x = 21$$

_____

$$2x^2 - 12x = -18$$

_____

$$2x^2 - 32x = -128$$

_____

Factoring - Solve for x

$$x^2 - 9x = -18$$

$$x^2 - 9x + 18 = 0$$
$$(x - 3)(x - 6) = 0$$
Set factors to equal 0
$$(3 - 3 = 0) \quad \& \quad (6 - 6 = 0)$$
$$x = 3 \text{ or } x = 6$$

$$x^2 - 8x = -16$$

$$x^2 - 8x + 16 = 0$$
$$(x - 4)(x - 4) = 0$$
Set factors to equal 0
$$(4 - 4 = 0) \quad \& \quad (4 - 4 = 0)$$
$$x = 4$$

$$x^2 - 2x = 63$$

$$x^2 - 2x - 63 = 0$$
$$(x - 9)(x + 7) = 0$$
Set factors to equal 0
$$(9 - 9 = 0) \quad \& \quad (-7 + 7 = 0)$$
$$x = 9 \text{ or } x = -7$$

Factoring - Solve for x

$$x^2 - 9x = -18$$

_____

_____

_____

_____

_____

_____

$$x^2 - 8x = -16$$

_____

_____

_____

_____

_____

_____

$$x^2 - 2x = 63$$

_____

_____

_____

_____

_____

_____

Factoring - Solve for x

$$x^2 + 16x = -63$$

$x^2 + 16x + 63 = 0$

$(x + 9)(x + 7) = 0$

Set factors to equal 0

$(-9 + 9 = 0)$ & $(-7 + 7 = 0)$

x = -9 or x = -7

$$x^2 - x = 42$$

$x^2 - x - 42 = 0$

$(x - 7)(x + 6) = 0$

Set factors to equal 0

$(7 - 7 = 0)$ & $(-6 + 6 = 0)$

x = 7 or x = -6

$$x^2 + 16x = 192$$

$x^2 + 16x - 192 = 0$

$(x - 8)(x + 24) = 0$

Set factors to equal 0

$(8 - 8 = 0)$ & $(-24 + 24 = 0)$

x = 8 or x = -24

Factoring - Solve for x

$$x^2 + 16x = -63$$

_____

$$x^2 - x = 42$$

_____

$$x^2 + 16x = 192$$

_____

Factoring - Solve for x

$$x^2 - 20x = -96$$

$x^2 - 20x + 96 = 0$

$(x - 12)(x - 8) = 0$

Set factors to equal 0

$(12 - 12 = 0)$ & $(8 - 8 = 0)$

x = 12 or x = 8

$$x^2 - 12x = -32$$

$x^2 - 12x + 32 = 0$

$(x - 4)(x - 8) = 0$

Set factors to equal 0

$(4 - 4 = 0)$ & $(8 - 8 = 0)$

x = 4 or x = 8

$$x^2 + x = 72$$

$x^2 + x - 72 = 0$

$(x + 9)(x - 8) = 0$

Set factors to equal 0

$(-9 + 9 = 0)$ & $(8 - 8 = 0)$

x = -9 or x = 8

Factoring - Solve for x

$$x^2 - 20x = -96$$

_____

$$x^2 - 12x = -32$$

_____

$$x^2 + x = 72$$

_____

Factoring - Solve for x

$$x^2 + 19x = 330$$

$x^2 + 19x - 330 = 0$

$(x - 11)(x + 30) = 0$

Set factors to equal 0

$(11 - 11 = 0)$ & $(-30 + 30 = 0)$

$x = 11$ or $x = -30$

$$x^2 + 40x = -300$$

$x^2 + 40x + 300 = 0$

$(x + 10)(x + 30) = 0$

Set factors to equal 0

$(-10 + 10 = 0)$ & $(-30 + 30 = 0)$

$x = -10$ or $x = -30$

$$x^2 + 28x = -160$$

$x^2 + 28x + 160 = 0$

$(x + 8)(x + 20) = 0$

Set factors to equal 0

$(-8 + 8 = 0)$ & $(-20 + 20 = 0)$

$x = -8$ or $x = -20$

Factoring - Solve for x

$$x^2 + 19x = 330$$

$$x^2 + 40x = -300$$

$$x^2 + 28x = -160$$

Factoring - Solve for x

$$x^2 - 18x = -81$$

$x^2 - 18x + 81 = 0$

$(x - 9)(x - 9) = 0$

Set factors to equal 0

$(9 - 9 = 0)$ & $(9 - 9 = 0)$

$x = 9$

$$x^2 - 3x = 180$$

$x^2 - 3x - 180 = 0$

$(x + 9)(x - 12) = 0$

Set factors to equal 0

$(-9 + 9 = 0)$ & $(12 - 12 = 0)$

$x = -9$ or $x = 12$

$$x^2 - 6x = 7$$

$x^2 - 6x - 7 = 0$

$(x + 1)(x - 7) = 0$

Set factors to equal 0

$(-1 + 1 = 0)$ & $(7 - 7 = 0)$

$x = -1$ or $x = 7$

Factoring - Solve for x

$$x^2 - 18x = -81$$

_____

_____

_____

_____

_____

_____

_____

$$x^2 - 3x = 180$$

_____

_____

_____

_____

_____

_____

_____

$$x^2 - 6x = 7$$

_____

_____

_____

_____

_____

_____

_____

Solve for y

$$2y - x = 4 + x + 3x$$
$$2y = 4 + x + 3x + x$$
$$2y = 4 + 5x$$
$$y = \frac{4 + 5x}{2}$$

$$7y - 2x = 5 + 2x + 3x$$
$$7y = 5 + 2x + 3x + 2x$$
$$7y = 5 + 7x$$
$$y = \frac{5 + 7x}{7}$$

$$4y + 2x = 2y + 14$$
$$4y - 2y + 2x = 14$$
$$2y = 14 - 2x$$
$$y = \frac{14 - 2x}{2}$$
$$y = 7 - x$$

$$x = \frac{y + 5}{13 - 3}$$
$$x = \frac{y + 5}{10}$$
$$y = 10x - 5$$

Solve for y

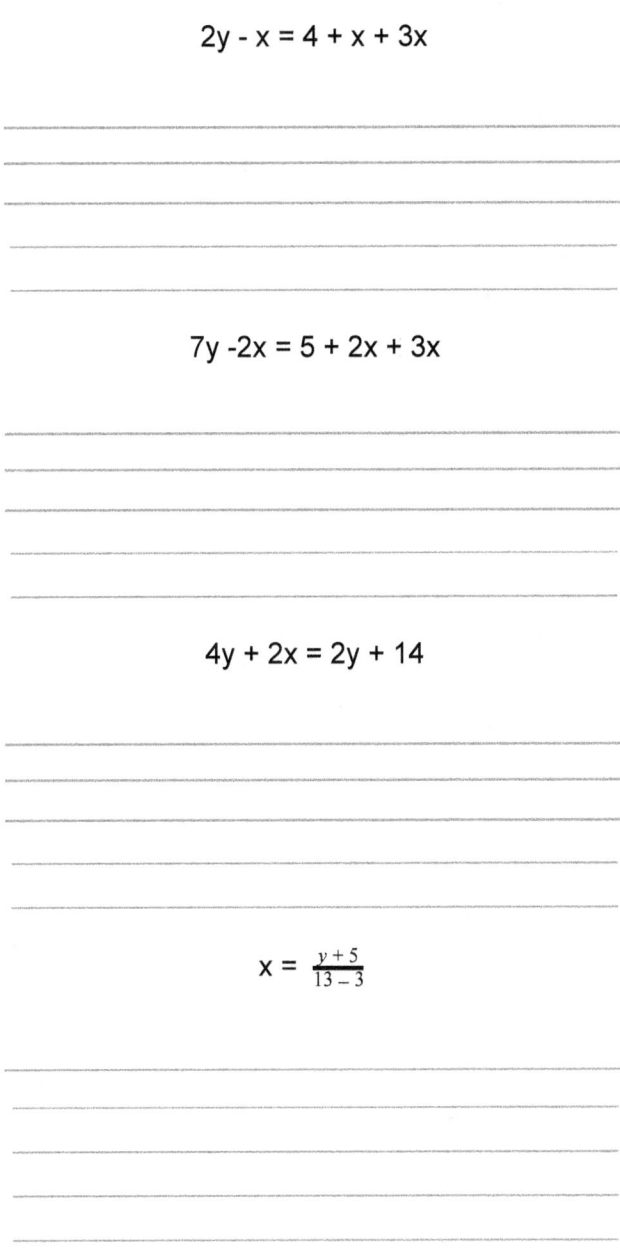

$$2y - x = 4 + x + 3x$$

$$7y - 2x = 5 + 2x + 3x$$

$$4y + 2x = 2y + 14$$

$$x = \frac{y + 5}{13 - 3}$$

Solve for y

$$3y + 2x = x - 2y + 5$$
$$3y + 2y = x - 2x + 5$$
$$5y = -1x + 5$$
$$y = \frac{-x+5}{5}$$
$$\text{or} \quad y = \frac{-1}{5}x + 1$$

$$5y + 3x = x - 3y + 2$$
$$5y + 3y = x - 3x + 2$$
$$8y = -2x + 2$$
$$y = \frac{-2x+2}{8}$$
$$y = \frac{-1x+1}{4}$$
$$\text{or} \quad y = \frac{-1}{4}x + \frac{1}{4}$$

$$4y + x = 7y(3) + 2x$$
$$4y + x = 21y + 2x$$
$$4y - 21y = 2x - x$$
$$-17y = 1x$$
$$y = \frac{x}{-17}$$
$$\text{or} \quad y = \frac{-1}{17}x$$

$$7y - 8x = 7y + 8x$$
$$7y - 7y = 8x + 8x$$
$$0 = 16x$$
$$y = \frac{16}{0}$$

Solve for y

$$3y + 2x = x - 2y + 5$$

_____

_____

_____

_____

_____

$$5y + 3x = x - 3y + 2$$

_____

_____

_____

_____

_____

$$4y + x = 7y(3) + 2x$$

_____

_____

_____

_____

_____

$$7y - 8x = 7y + 8x$$

_____

_____

_____

_____

_____

Solve for y

$$y + 3x = 2y + 15$$

$$
\begin{aligned}
y + 3x - 2y &= 15 \\
y - 2y &= 15 - 3x \\
-1y &= 15 - 3x \\
y &= \frac{15 - 3x}{-1} \\
y &= -15 + 3x
\end{aligned}
$$

$$2x - 3y = y + x - 3$$

$$
\begin{aligned}
2x - 3y - y &= x - 3 \\
-3y - y &= x - 3 - 2x \\
-4y &= -1x - 3 \\
y &= \frac{-1x - 3}{-4} \\
y &= \frac{x}{4} + \frac{3}{4}
\end{aligned}
$$

$$4x + 5 = 3y - 8$$

$$
\begin{aligned}
-3y + 4x + 5 &= -8 \\
-3y + 4x &= -8 - 5 \\
-3y &= -4x - 13 \\
y &= \frac{-4x - 13}{-3} \\
y &= \frac{4x}{3} + \frac{13}{3}
\end{aligned}
$$

$$y + 12 = y - x$$

$$
\begin{aligned}
y + 12 - y &= -x \\
y - y &= -x - 12 \\
0y &= -x - 12 \\
y &= \frac{-x - 12}{0}
\end{aligned}
$$

Solve for y

$$y + 3x = 2y + 15$$

$$2x - 3y = y + x - 3$$

$$4x + 5 = 3y - 8$$

$$y + 12 = y - x$$

Solve for y

$$28 = 12(\tfrac{y}{9}) + 4$$
$$-12(\tfrac{y}{9}) = 4 - 28$$
$$-12(\tfrac{y}{9}) = -24$$
$$\tfrac{y}{9} = \tfrac{-24}{-12}$$
$$\tfrac{y}{9} = 2$$
$$y = (2)(9)$$
$$y = 18$$

$$\tfrac{y+5}{3} 4 = 12$$
$$\tfrac{y+5}{3} = \tfrac{12}{4}$$
$$\tfrac{y+5}{3} = 3$$
$$y + 5 = (3)(3)$$
$$y + 5 = 9$$
$$y = 9 - 5$$
$$y = 4$$

$$3(\tfrac{15}{y}) - 3 = 6$$
$$3(\tfrac{15}{y}) = 6 + 3$$
$$3(\tfrac{15}{y}) = 9$$
$$\tfrac{15}{y} = \tfrac{9}{3}$$
$$\tfrac{15}{y} = 3$$
$$y = \tfrac{15}{3}$$
$$y = 5$$

Solve for y

$$28 = 12\left(\frac{y}{9}\right) + 4$$

$$\frac{y+5}{3}\,4 = 12$$

$$3\left(\frac{15}{y}\right) - 3 = 6$$

Solve for y

$$12y = 3^3 + 2y$$

$$
\begin{aligned}
12y &= 27 + 2y \\
12y - 2y &= 27 \\
10y &= 27 \\
y &= \frac{27}{10} \\
y &= 2\frac{7}{10}
\end{aligned}
$$

$$2(y - 4) = 3y + 8$$

$$
\begin{aligned}
2y - 8 &= 3y + 8 \\
2y - 8 - 3y &= 8 \\
2y - 3y &= 8 + 8 \\
-1y &= 16 \\
y &= \frac{16}{-1} \\
y &= -16
\end{aligned}
$$

$$3(y + 5) = 6y - 9$$

$$
\begin{aligned}
3y + 15 &= 6y - 9 \\
3y + 15 - 6y &= -9 \\
3y - 6y &= -9 - 15 \\
-3y &= -24 \\
y &= \frac{-24}{-3} \\
y &= 8
\end{aligned}
$$

$$y(-3 - 9) = 27 - 3y$$

$$
\begin{aligned}
-3y - 9y &= 27 - 3y \\
-3y - 9y + 3y &= 27 \\
-9y &= 27 \\
y &= \frac{27}{-9} \\
y &= -3
\end{aligned}
$$

Solve for y

$$12y = 3^3 + 2y$$

$$2(y - 4) = 3y + 8$$

$$3(y + 5) = 6y - 9$$

$$y(-3 - 9) = 27 - 3y$$

Working with negative numbers

Add/Subtract

Like signs - Add	5	−5
	$\underline{+3}$	$\underline{+-3}$
	8	−8
Unlike signs - Subtract	5	−5
	$\underline{-3}$	$\underline{+3}$
	2	−2

Multiply/Divide

Like signs - Positive	5	−5
	$\underline{\times\ 3}$	$\underline{\times\ -3}$
	15	15
Unlike signs - Negative	5	−5
	$\underline{\times -3}$	$\underline{\times\ 3}$
	−15	−15

Algebra rules for arithmetic

$$a(b + c) = ab + ac$$

$$a\left(\frac{b}{c}\right) = \frac{ab}{c}$$

$$\frac{a}{b} + \frac{c}{d} = \frac{ad + bc}{bd}$$

$$\frac{a + b}{c} = \frac{a}{c} + \frac{b}{x}$$

$$\frac{ac + bc}{c} = a + b$$

Exponents

$$a^m a^n = a^{m+n}$$

$$(a^m)^n = a^{mn}$$

$$(ab)^n = a^n b^n$$

$$a^{-n} = \frac{1}{a^n}$$

$$\left(\frac{a}{b}\right)^{-n} = \left(\frac{b}{a}\right)^n$$

$$\frac{a^n}{a^m} = a^{n-m}$$

$$a^0 = 1$$

Other books from Timothy Schablin Mathematics

equals(me)
Pre-Algebra Practice

equals(me)
Algebra Practice

equals(me)
Radicals Practice

Fun with PEMDAS
PEMDAS Practice

Available at Amazon Books or

https://timothyschablin.wixsite.com/equalsme

About the author

Timothy Schablin is a graduate of the Hutchinson Technical College where he studied algebra, trigonometry, physics, mathematical techniques, and technical related fields. He also holds two certificates of physics from Davidson College, AP Physics I & AP Physics II: Challenging Concepts.

Timothy Schablin tutors math to 5th, 6th, 7th, and 8th graders at a local middle school. He is also a member of Minnesota MathCorps and has authored mathematical software.

Besides studying & tutoring math and physics, Timothy enjoys astronomy. He spends vacation time canoeing the Minnesota River bottom.